I0605536

union
square
kids
NEW YORK

UNION SQUARE KIDS and the distinctive Union Square Kids logo are trademarks of Hachette Book Group, Inc.

ISBN 978-1-4549-6170-3

Library of Congress Control Number: 2025024835

Union Square Kids books may be purchased in bulk for business, educational, or promotional use. For more information, please contact your local bookseller or the Hachette Book Group's Special Markets department at special.markets@hbgusa.com.

Printed in Guangdong, China

Lot #:

2 4 6 8 10 9 7 5 3 1

08/25

unionsquareandco.com

Text by Heather Alexander
Book design by Clarisse Hassan
Edited by Sarah Carpenter

For all my animal-loving friends.—H.A.

For Elliott, my own little boy who loves animals.
May you forever stay curious.—S.W.

5 Minute Genius Stories

JANE GOODALL

Written by Heather Alexander
Illustrated by Sally Walker

How to use this book

In this book you'll find **ten genius stories** to read, each one just **5 minutes long**.

At the end of each story, explore an informative **"all about"** spread.

Want to learn more? Turn to the back of the book to discover a **timeline of key events**.

union square kids
NEW YORK

WHO IS JANE GOODALL?

Dr. Jane Goodall's discoveries changed our understanding of chimpanzees in the wild, and showed the world why these amazing creatures need to be protected. Follow her life as a trail-blazing scientist and animal activist.

WHICH 5-MINUTE GENIUS STORY WILL YOU READ TODAY?

YOUNG SCIENTIST

The Child Who Noticed What Others Missed

Jane Goodall was a curious and patient girl. Above all, she was fascinated by animals.

On her first birthday, Jane's father gave her a large, stuffed chimpanzee with soft brown fur. It was named Jubilee and immediately became Jane's favorite toy.

Jane and Jubilee played together in the garden.

They squatted together by tide pools.

They snuggled together in bed.

Young Jane would leave breadcrumbs on her windowsill. Each day, a robin would visit to eat them. One day, it flew inside and made a nest in Jane's bookcase! She watched quietly in fascination.

Jane had a younger sister named Judy. As they grew, the girls became best friends. They played games and listened to their mother read stories.

Jane liked the books about animals the best, but she longed to see them up close. She wanted to understand why animals did what they did.

Jane was filled with many questions.

Why do bumblebees buzz around flowers?

Why does a dog turn in circles before settling down to sleep?

How do eggs come out of hens?

Five-year-old Jane decided to discover the answer to that last question herself.

One day Jane snuck into the little wooden henhouse in their backyard. She nestled beside a bale of hay and waited.

After a while, a hen strutted in and sat on its nest. It squawked, warning her away. Jane didn't go. She stayed as still and silent as a statue.

Hours passed. Scratchy straw tickled Jane's legs, but she willed herself not to squirm. The hen grew used to her being there.

Shadows appeared, and still Jane waited. Finally, her patience paid off. Something was happening!

Jane watched wide-eyed as the hen ruffled its feathers and rose up a bit. With a plop, an egg dropped from between its legs into the softness of the nest!

Letting out a whoop, Jane raced outside to tell her mom.

She was surprised to find the sky now dark. She heard her name being called. Why were the neighbors in her front yard?

Her mom had been frantic with worry. Jane had been missing for hours! She'd looked everywhere and even called the police. No one had thought to search the henhouse.

Now her mom listened as Jane described the wonder of the egg being laid. She told Jane she had the qualities of a great scientist—she was curious, asked questions, and wasn't afraid to search out answers.

Jane was so excited. She'd made her first scientific observation!

How to Observe Nature Like JANE

As a scientist, Jane used the power of observation in the famous discoveries she'd later make with chimpanzees. Observation is watching something or someone carefully in order to learn more about them.

Start by choosing a safe location and an animal you can **observe** nearby, such as your pets at home, a squirrel on a tree, an ant in the dirt, or a turtle in a stream.

Focus your attention on your chosen animal. Stay very still, so the animal isn't scared away. **Be patient** like Jane.

Day: May 5 Time: 10 AM
Weather: Warm and Sunny
Location: Park pond

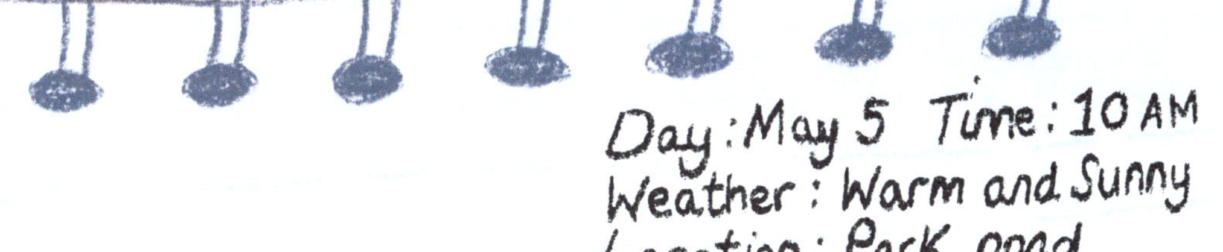

Keep a **field journal**. In it, record your location, day and time, weather, and what you observe. Use your five senses to gather information. Note other creatures and plants nearby.

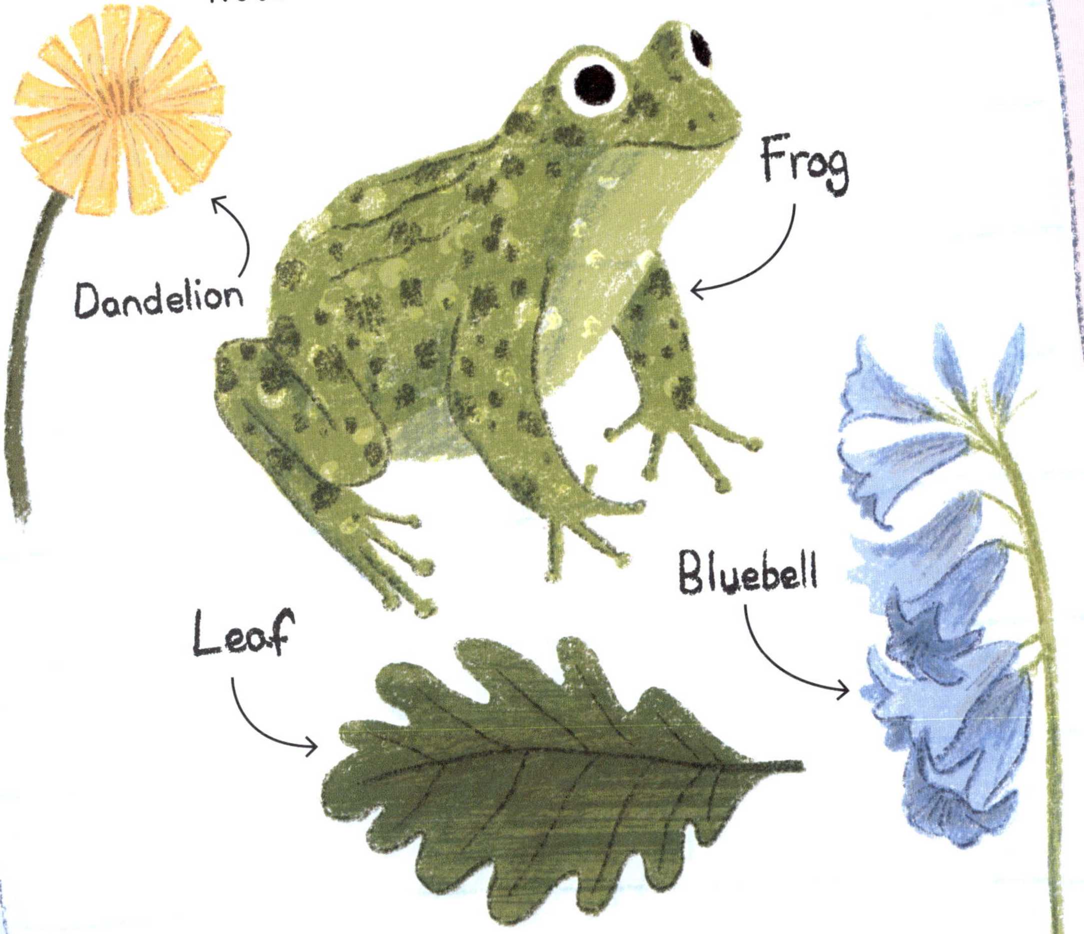
Dandelion
Frog
Leaf
Bluebell

Try returning to the same spot on other days to **record** what has changed.

ANIMAL WHISPERER

The Girl Who Talked to Animals

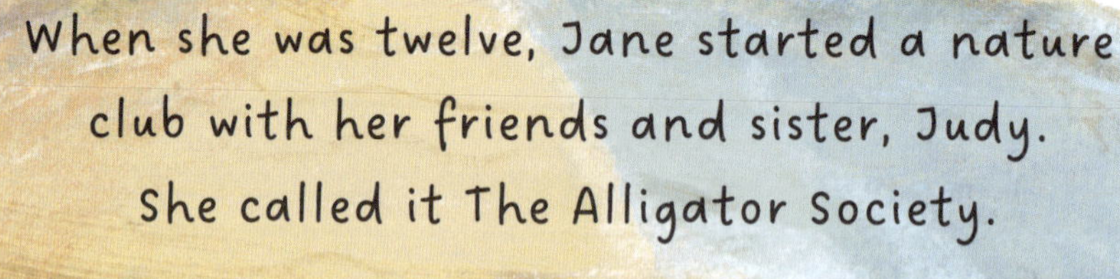

When she was twelve, Jane started a nature club with her friends and sister, Judy. She called it The Alligator Society.

Sue was Ladybird,

Judy was Trout,

Sally was Puffin . . .

. . . and Jane was Red Admiral. She led them on nature walks. The girls took notes and sketched pictures.

They published them in their own homemade magazine that they shared with their family and friends.

They also created a museum to display their finds. Jane learned lots about nature.

Jane realized she could learn something new from every animal.

Rusty was a black dog that lived at the hotel around the corner from Jane's house. Every morning, Rusty trotted to Jane's door and barked loudly to announce his arrival.

Jane taught Rusty how to jump through a hoop,

shake paws,

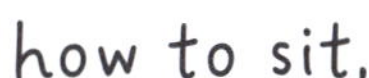

how to sit,

and even how to close a door.

Teaching Rusty tricks took an extremely long time, but Jane never got angry or raised her voice. She used lots of love and plenty of cuddles as rewards.

Around the 1940s, scientists believed that animals were very different from humans—that they didn't think or act like humans.

But Jane noticed that Rusty missed her when she was gone . . .

. . . and he was joyful when she returned.

Rusty felt bad when he did something wrong and was confused when he lost sight of a tossed ball.

Jane knew that Rusty shared many of the same emotions that she did.

She realized that just because everyone believes something, it isn't necessarily correct.

When Jane wasn't playing with Rusty or on nature walks, she could be found up in her favorite beech tree in her family's garden. She named it Beech.

She perched in the branches of Beech and read.

Over and over again, Jane read *The Story of Doctor Dolittle*, by Hugh Lofting and *Tarzan of the Apes*, by Edgar Rice Burroughs.

Doctor Dolittle was a country doctor who could talk with animals. Tarzan was a boy raised by apes in faraway Africa.

Jane thought that animals were smarter than people realized. Some day, she would travel to Africa and spend time with animals in the wild herself.

How to Talk to a CHIMP

Chimps use many sounds to communicate. They can't speak words like humans, because their vocal cords are different from ours. Jane would later learn to communicate with chimps, much like Doctor Dolittle.

Chimps will greet one another with loud **"pant-hoots"** and **"hoos."** Each chimp has its own recognizable pant-hoot.

When eating something yummy, they will **grunt**.

To warn of danger or show anger, they will **scream or bark**.

They'll blow raspberries when happy and **laugh** when playing.

BOLD AMBITIONS

The Young Woman Who Dared to Dream

Young Jane dreamed of becoming a scientist and going to Africa to study animals. But when she told people, they laughed and said it would never happen.

Around the 1950s, the jobs available to women in the United Kingdom were in professions—they were secretaries, nurses, or teachers. Few women were scientists, and the ones who were had college degrees.

Jane received top grades, but her family couldn't afford tuition to send her to college. Also, girls were discouraged from traveling to faraway places, especially on their own.

Nevertheless Jane's mother wholeheartedly supported her. She insisted her daughter never give up on her dreams, even if they took time and hard work to achieve. She wanted Jane to be bold.

After high school, Jane found a job as a secretary. She hated being inside an office building.

One day, Jane received a letter from her school friend, Clo, who now lived in Kenya, a country in East Africa. She invited Jane for a visit.

Jane was determined to go. Round-trip boat tickets were expensive, so Jane moved back home and took a job as a waitress. Everything she earned, she slipped under the parlor carpet.

Eventually, Jane bought a ticket and set sail to Africa. The journey lasted three weeks. Then there was a two-day train ride to the capital city of Nairobi.

Clo and her family welcomed Jane at the station.

On the drive to Clo's family's farm, Jane couldn't believe her eyes. A giraffe stood by the side of the dirt road. An aardvark crossed in front of them. Wow!

Jane loved Kenya even more than she'd dreamed. She decided to stay. Africa was a huge continent with so much more to experience.

At a party, Jane told her new friends that she hoped to work with animals. They said she must meet Dr. Louis Leakey—right away!

Dr. Leakey was a scientist who studied animal behavior. He was especially interested in great apes.

Dr. Leakey was researching chimpanzees. He hoped to scientifically prove that chimpanzees and our human ancestors had much in common. In the 1950s, this was a radical thought.

Dr. Leakey hired Jane as a secretary. He was so impressed by her knowledge that soon she was working on fossil digs.

Dr. Leakey had heard of a troop of chimps living in the rainforest in Tanzania. He was trying to find a scientist willing to study them up close. No one had ever done this in the wild before, but he thought Jane would be perfect.

Jane finally achieved her goal: she would have the chance to study chimps in the wild in Africa!

Bold RESEARCH

Apes are humans' closest living biological relatives. Dr. Louis Leakey and Jane Goodall's research with chimpanzees paved the way for humans to understand more about themselves and other animals.

Apes are mammals that are part of the **primate family**. Humans are also primates.

Monkeys are also primates, but they are **not apes**. Monkeys tend to be smaller than apes, and most monkeys have tails, while apes don't.

Chimps are are found in **Africa**. They are more closely related to humans than any other great ape.

We now think that humans, chimps, and bonobos all evolved from a **single ancestor**, six or seven million years ago.

PATIENCE AND PERSEVERANCE

Determination Wins the Day

When Jane was around 25 years old, Dr. Louis Leakey agreed that she could help him with his research by going into the field and studying chimps in the wild.

Even though Jane didn't have formal training, Dr. Leakey saw that she had more important qualifications for the job.

He had seen in the office and out in the field on fossil digs that she was a hard worker.

Dr. Leakey was most impressed by Jane's patience and perseverance. She kept trying and wouldn't give up—even when a job wasn't easy.

But the government officials refused to allow a young woman to travel alone to such a remote area. Jane was determined to find someone to go with her. But who?

Jane asked her mom, who said yes! She didn't want Jane to miss out on this perfect opportunity.

They arrived at the Gombe National Park on the eastern shore of Lake Tanganyika in the summer of 1960. With the help of two scouts, they set up a large tent in a clearing by a small stream.

They covered the front with mosquito netting. Nearby, they dug a big hole to be used as a toilet.

There was no running water or electricity. But that didn't worry Jane or her mom.

Inside the tent were cots and an area for washing up. A cook named Dominic would also live here.

Right away, Jane set off to have a look around.

Jane hiked through trees heavy with broad leaves. As she pushed through tangled vines and crawled in spiky grass, snakes slithered by her feet.

Bush pigs, vervet monkeys, leopards, hyenas, and mongoose lurked nearby. Swarms of tsetse flies circled.

In the distance, a chimp hooted.

In the morning, Jane climbed a peak and peered through her binoculars.

She didn't spot a single chimp anywhere. She scanned the horizon all day.

At dawn the next morning, she returned to the peak. Nothing.

But Jane knew they were there. The chimps' hooting calls echoed throughout the forest. Were they hiding?

Over the next eight weeks, Jane searched, even in the pouring rain and the dark of night. She walked great distances through the valleys and up mountains. Still no chimps. But Jane wouldn't give up.

More months passed by. Jane learned to walk silently through the underbrush. Her body adjusted to the heat, and the bug bites became less itchy. But where were the chimps?

Was she doing something wrong? Jane didn't want to write Dr. Leakey to tell him she'd failed. So she stayed patient.

Then, one day on the peak, Jane caught sight of a group of chimpanzees in the distance.

The next day she saw more! Finally, her perseverance had paid off.

HANGING IN There!

While it was difficult for Jane to travel through the dense jungle, she observed that the chimps traveled through their environment with ease.

On the ground, **chimps walk** on all fours. They use their feet and the knuckles of their hands for support.

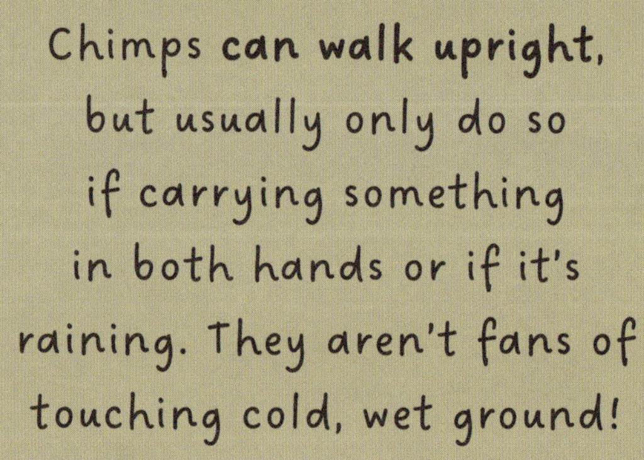

Chimps **can walk upright**, but usually only do so if carrying something in both hands or if it's raining. They aren't fans of touching cold, wet ground!

To move through the tree tops, **chimps swing** from branch to branch, hanging from their hands. Their arms are longer and stronger than ours, and can rotate in all directions from the shoulder.

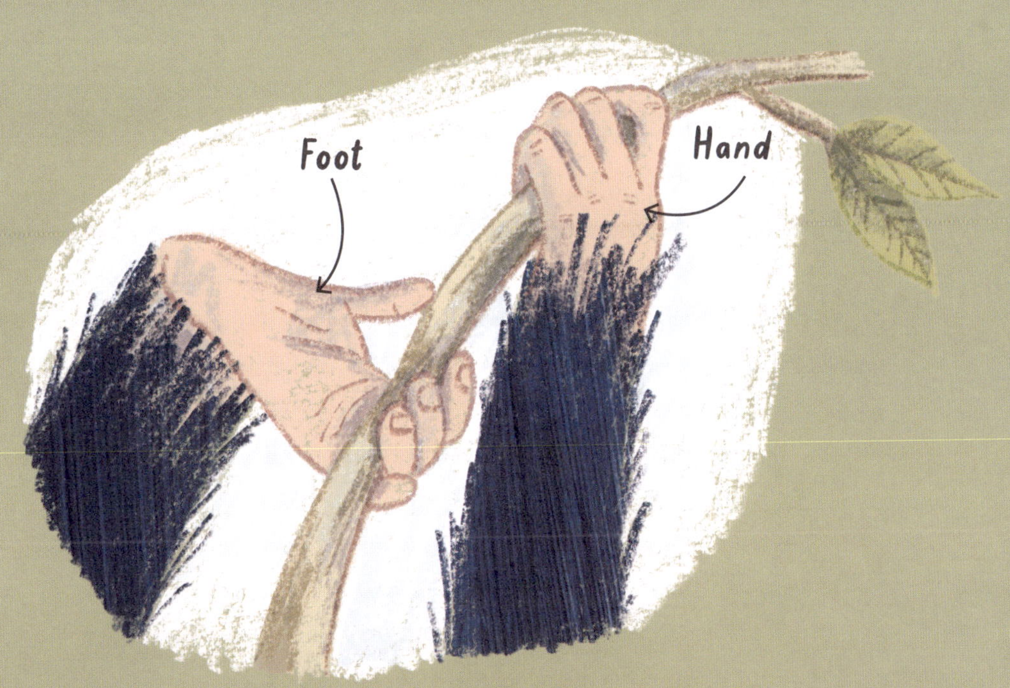

Chimps have **thumbs** on their hands and flexible big toes on their feet that act like thumbs. This allows their feet to grip branches and pick things up. It's like having four hands.

CLOSER THAN EVER

Jane Gets to Know the Chimps

In 1960, when Jane began studying chimps in Africa, she observed their movements from a distance. At first, the chimps stayed far away from her. But Jane understood. She was new to the chimps' world.

It would take time for them to adjust to Jane's arrival and trust her.

But from Jane's usual spot on the Peak, it was tricky to see or understand what they were doing. She grew frustrated. She had to get closer.

No scientist had ever dared approach a wild chimp before. How should she do it?

Jane dressed head-to-toe in tan and drab green to blend in with the surroundings.

She climbed a tree near where a troop of chimps liked to gather. Jane was careful to not make any loud sounds or sudden movements, but she didn't hide, either. She wanted to show she meant no harm.

Jane pretended she was a chimp too. She scratched herself and gobbled berries.

She noticed that it upset the chimps to be stared at directly, so she acted as if she didn't care what they did.

At first, the chimps fled. But as months went by, they grew less fearful and more curious. Jane sensed them watching her watching them.

In a little notebook, Jane wrote down what the chimps ate, how they played, and when they fought.

The more she watched, the more she learned.

She saw how, each night, the chimps bent branches and used leaves to construct sturdy beds high in the branches to sleep.

Jane wondered how it would feel to sleep in a cozy chimp nest.

Gradually, Jane began to recognize the different chimps. She gave them names.

David Greybeard was a handsome male with silver hair on his chin. He was determined and gentle.

Goliath was the most powerful male in the troop. He was adventurous and good friends with David Greybeard.

Mr. McGregor was bald on his head and shoulders and could be grumpy. He would shake a branch at Jane if she came too close.

Flo was the oldest female and very popular. She had a big bulbous nose and a notch in one ear.

Fifi was Flo's baby. She often rode on her mother's back. Faben and Figan were Flo's sons.

Giving the chimps names shocked many scientists. Back then, it was believed that animals being studied should be assigned a number.

But to Jane, the chimps were more than just numbers. She had come to know each one's unique personality, and she felt they deserved names.

Chimps UP CLOSE

Chimpanzees are roughly the size of a small human, but about twice as strong. A grown male chimp stands about 4-5 feet tall and weighs about 100 pounds. Females are a bit smaller.

A group of chimps is called a **troop**.

Every chimp is different. **Just like humans**, they vary in height, body shape, and personality.

DAVID GREYBEARD

The Chimp Who Took a Chance

Jane spent a long time quietly observing the chimps.

Chimpanzees are strong, wild animals, and interacting with them could be dangerous.

But over time, Jane got to know one chimp in particular better than all the others.

One day, a brave male chimp wandered to the edge of Jane's camp. Dominic the cook had seen him there the day before, munching the ripe, red fruit of a nearby palm nut tree. He had come back for more.

Jane recognized the chimp by the gray hairs on his chin. It was the one she and her team called David Greybeard.

As David Greybeard ate, he caught sight of a bunch of bananas that was part of Jane's food supply.

He ran over, swiped the bananas, then disappeared into the forest!

The next day Jane left out another bunch of bananas. Would he come back?

Finally, in the late afternoon, Jane spotted a dark furry shape behind a tree across the clearing.

It was David Greybeard, gobbling palm nuts again. He kept a close watch on Jane. She stayed very still. When the palm nuts were all gone, David Greybeard stole the bananas and raced off.

For several more days, the same thing happened. Jane stayed patient, and David Greybeard grew bolder. Now, as he entered the camp, he went right for the bananas.

One day, Jane tried an experiment. She placed a banana in her open palm.

How would David Greybeard react?

The chimp eyed the banana.

Then he nervously moved toward her.

One step, then another.

Soon he stood beside her.
Jane slowly stretched out her hand.

David Greybeard took the banana gently, without snatching it. This was a huge deal! It was the first time in known history that a human had interacted with a wild chimpanzee in this way.

Over time, David Greybeard grew to trust Jane.

He would come to sit beside her, greeting her with a soft, "Hoo!"

David's two closest friends, Goliath and William, watched shyly from a distance as Jane brought David bananas.

After many months, chimps of all ages were gathering around Jane.

The chimps had decided that if David Greybeard trusted Jane, then they would too.

SOCIAL Animals

Chimps live in large groups called communities. Communities have between fifteen and eighty males and females.

Within a **community**, you'll find families, good friends, and some chimps who don't like one another.

The chimps will split into smaller groups, called **parties**, to play, eat, and travel.

The highest-ranking male chimp in a community is called the **alpha**.

Sometimes chimps fight to be the **alpha**, but the biggest or the strongest doesn't always become the boss. It can be the smartest chimp, or the one with the most friends. The alpha's close male friends help him protect the other chimps.

Although a chimp community is ruled by males, there is always a **high-ranking female**.

High-ranking female

She's often the **mother** of many babies or excellent at finding food.

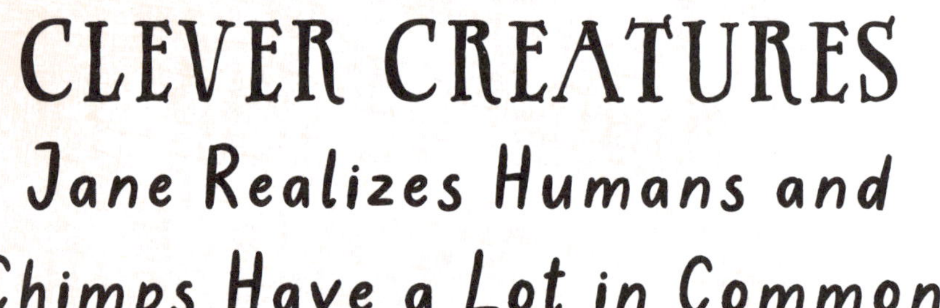

CLEVER CREATURES

Jane Realizes Humans and Chimps Have a Lot in Common

Jane discovered that chimps are very intelligent—and use gestures to express their feelings.

If a chimp wants more food, it will extend its palm.

If frightened, it will grin widely with teeth showing.

When angry, a chimp's hair stands up. It might wave its arms and throw rocks or sticks.

When happy to see another chimp, chimps give hugs or kisses.

A lower-ranking chimp will bow to greet a higher-ranking chimp.

The chimps displayed many of the same emotions Jane felt—happiness, sadness, fear, anger, grief, and jealousy.

Jane saw that the chimps formed lasting friendships. Best friends held hands, sat next to each other, and shared food.

Chimp friends spent hours grooming each other. They combed their fingers gently through their friend's long fur, quietly picking out specks of dirt, grass, and ticks.

When the other chimp felt it was their turn, they gave their friend a poke!

David Greybeard allowed his new friend Jane to groom his fur. Later on, Jane decided to stop this. It was wrong for humans to interfere in natural chimp behavior.

However, Jane's biggest—and most famous—discovery was about to come.

One day loud pant-hoots caught Jane's attention. She followed the calls and found David Greybeard sitting in front of a termite mound. What was he up to?

David had broken off a long grass stalk and was pushing it into the mound through a hole on top.

The termites, angry about a pointy thing invading their nest, tried to attack the stick and grabbed on. Whoosh! David pulled up the stalk and slurped the plump insects. Tasty!

Jane was amazed. David Greybeard was using the grass as a tool! Before this, scientists had assumed humans were the only animals intelligent enough to make and use tools.

Jane also watched David Greybeard and his friend Goliath strip leaves from twigs and break off the ends to make their termite-fishing tools work better.

She watched chimps use sticks to pull sweet honey from a hive.

The chimps would crumple leaves to soak up water from a hard-to-reach place, then squeeze them like sponges into their mouths.

Jane saw them use leaves to scoop water to drink and wipe dirt from their bodies.

Chimps were resourceful and much more intelligent than scientists had believed.

Chimps are SMART

Chimpanzees have the largest brains relative to body size of any primates, other than humans.

They can **learn** and solve problems.

They use **tools**, and can remember where they have been so they can return to where they once found food.

They will **watch** another chimp's face and body to determine how it is feeling.

They **communicate** with one another and display different personalities and emotions.

LOVE IN OUR NATURE

Jane Learns Some Lessons in Love

National Geographic magazine heard about Jane working with chimps in Africa and sent a photographer named Hugo van Lawick to take pictures of her. Jane was a little nervous. What if Hugo got in the way?

Hugo was respectful, and the chimps were okay as long as Jane stayed nearby. He took pictures while Jane took notes.

Hugo loved being with the chimps. He decided to stay on and work side by side with Jane.

Over the next few years, Jane and Hugo fell in love, got married, and had a son. They nicknamed him "Grub."

Jane had noticed that many of the female chimps were good mothers. Once again, Jane saw that humans and chimps were more similar than people might have thought.

In 1964, the chimp Jane had named Flo gave birth to a son. Jane named him Flint. Flo cradled Flint, giving him a lot of love, attention, and hugs. She tickled Flint to make him smile and always protected him from danger.

A newborn chimp is as helpless as a human baby. Babies sleep snuggled in a nest with their mom. A mother chimp cares for her baby by herself. Adult male chimps don't help raise their young.

For the first five months, a baby chimp clings to its mom. With strong fingers, it holds tight to her belly fur as she walks.

Later, the baby rides atop its mom's back.

When it is about three or four years old, the baby starts to walk unsteadily and begins to play with other baby chimps. The young chimps tumble, chase, and roll together.

Young chimps stay close to their mom for the first seven years. They learn by observing and imitating.

Flo gave Flint encouraging pats if he needed reassurance. She distracted him away from bad or unsafe behavior. She helped him practice climbing trees—and was ready for a rescue mission when he got stuck!

Mothers show their children where to forage for the best food. To learn to eat, chimp babies taste the food from mom's mouth.

Older siblings help out too. Flint's four-year-old sister Fifi loved to groom him and give him rides on her back. She grew up to be an excellent mother.

Flint's older brothers Faben and Figan played tug-of-war with him. They were less gentle!

Over the decades that she was at Gombe, Jane watched Flo's babies have babies, and those babies have their own babies.

Grub grew up and moved to Tanzania—and had kids of his own.

Jane's grandchildren learned to love nature, too!

Learning from MOM

Chimps learn what foods to eat and where to find it from their mothers.

At first, a baby chimp feeds on its **mother's milk**.

Over time, they watch Mom eating **fruits, leaves, flowers, nuts, and seeds.** They try some, too!

Chimps watch their mothers and learn to **bang nuts on a stone** to crack the tough shells.

Chimps will also **eat meat** and will share it with others in their group.

Chimps learn from their elders how to **hunt small animals**, and will work together to catch something to eat.

NATURAL HABITAT

Jane Fights for her Home

While studying chimpanzees in Africa, Jane made ground-breaking discoveries. She would soon appear on the covers of magazines and become a global superstar. Everyone would want to know more about her and the chimps.

Jane was excited to share her discoveries, though she preferred to stay out of the spotlight.

But Dr. Leakey worried that the scientific community wouldn't take Jane seriously, and that they would say she wasn't a "real" scientist. He thought it would help if she had a university degree.

So Jane bid a tearful goodbye to David Greybeard, Flo, and the other chimps. She hated to leave her forest home, but she knew the education would make her more useful to them.

In 1962, Jane traveled to Cambridge University in England to begin her studies. Jane didn't feel at home in this new environment at first.

Sometimes she found being in a classroom frustrating, especially when professors tried to teach her the "right way" to interact with animals.

In the end, Jane learned from them, and they learned from her.

Jane recieved her PhD in 1966 and became Dr. Jane Goodall. She was now officially a primatologist, or a scientist who studies primates.

When Jane returned to Gombe, she set up a proper research center and hired graduate students and local field scouts. She immediately felt at home again.

Jane was committed to learning everything about the chimps. But as decades passed, Jane began to notice something very scary.

The chimps were losing their natural habitat. The changing environment meant there were fewer chimps in the forest.

Logging companies were cutting down the tall trees for wood and charcoal. More trees were knocked down to create roads for the machines to drive on, and to build villages and farms.

Without the trees, the chimps had no place to sleep. They had little fruit and leaves to eat. They were forced to change their behavior or move. Many did not survive.

And that wasn't all. The humans in the villages passed on diseases that sickened the chimps, who hadn't had contact with humans before. New roads also made it easier for hunters—called poachers—to drive into the forest and kill the chimps.

Chimps were endangered and now in serious danger of becoming extinct. Jane couldn't imagine an Africa without chimpanzees. But what could she—just one person—do?

A lot! She had fought as a girl and as a young woman to come to Africa and study the chimps. Now she would fight to save them.

Jane's passion and her voice were her great power. She became an activist. She traveled the world to share the story of the vanishing chimps.

Thanks to the work of Jane and others, chimpanzees became a protected species, and some of their habitat would be protected, too.

At Home in the TREES

Chimps are arboreal, meaning they live among the trees. They often spend all night and about half of their day up in the trees.

Chimps build their beds, called **nests**, in the canopies to avoid nighttime predators, like leopards. They drum on tree trunks to communicate. Fruits, leaves, and nuts make up most of a chimp's diet.

Just as chimps rely on trees, trees rely on their furry house guests. Chimps are fruit-eaters. When they poop, they **disperse fruit seeds** onto the ground. The seeds take root and sprout into new trees.

If chimps are forced to leave an area, fewer trees are able to grow. This causes **problems for humans** as well as the animals that live in them. Trees help clean the air we breathe and help the planet to stay at the right temperature.

JANE'S LEGACY

Activist, Environmentalist, and Leader

Throughout her life, Jane fought for chimps and the planet. In 2024, she celebrated her 90th birthday with friends, colleagues, and family.

The birthday party was held on a beach in California. There was a cake with ninety blazing candles. But, even better, there were ninety dogs!

Dog after dog filed past the woman who had loved animals since the moment she was born. The dogs were, by far, Jane's favorite party guests!

As Jane bent to greet each one, she remembered her childhood dog, Rusty, and all he'd taught her.

Over her long lifetime, Jane has taught the world so much about animals. Her groundbreaking discoveries proved how similar we are to our closest living relative.

We learned that chimpanzees have personalities and emotions. They have loving families, friends, and communities.

They use tools, solve problems, and are intelligent.

People had once thought that humans were superior to animals. But Jane helped us understand that humans are part of the animal kingdom, not separate from it.

She made people realize how important it is to protect not only wild chimps but also ones kept in captivity.

Jane rallied against chimps being housed in small cages in labs, and mistreated. She visited zoos and tried to improve conditions, showing how enclosures could better mimic homes in the wild.

In her later years, Jane turned her attention to the health of the planet. She pushed to have trees planted to rebuild habitats. Through one of her many programs, more than one million trees were planted.

She raised money for and built sanctuaries where animals could live safely on protected land.

In 2025, Jane receieved the Presidential Medal of Freedom, given to amazing people who make our world better. It was given to her by President Joe Biden.

Jane didn't achieve all this alone. Her work continues with each new generation that joins her efforts to protect the natural world.

Many people are working to find ways to support the environment, to ensure there will continue to be thick forests, clean water, and chimpanzees swinging through the trees.

Just like the dogs on the beach, it's easy to imagine the generations of chimps in Gombe lining up for the chance to thank Jane Goodall.

She had the patience to observe and understand them. She cared so deeply and loved so much, and she did everything she could to protect them.

Jane's stuffed chimp, Jubilee, still sat on the top of her dresser many decades after he was given to her. The toy was worn and threadbare, but it reminded Jane of her lasting love of animals.

Jane spent a lifetime protecting animals, and now younger generations are taking up her example and protecting them, too.

Following JANE'S EXAMPLE

Small actions can add up to make big differences. Jane was someone who watched and learned, then she made discoveries and took action.

By **observing** animals, you can learn more about how to protect and help them.

Jane's organization for kids, Roots & Shoots, focuses on animals and ways to protect them. You can **visit the site** at rootsandshoots.org

On her own website, Jane wrote:

"If each of us does our part, all the pieces of the puzzle come together and the world is a better place because of you."

JANE GOODALL'S JOURNEY TO BECOMING AN ENVIRONMENTAL CHAMPION

1934

April 3

Jane is **born** in England.

1939

Hidden in a henhouse, Jane makes her first **scientific discovery.**

1946

Jane starts **The Alligator Society** with her friends.

1963

National Geographic publishes the **first cover story** about Jane's research.

1961

David Greybeard comes to visit. Jane discovers that chimps make and use tools, communicate with one another, and eat meat.

1966

Jane earns her Ph.D. from **Cambridge University**.

1967

Jane **gives birth** to her son, "Grub."

1977

The Jane Goodall Institute is founded.

2025

In 2025, Jane is awarded the **Presidential Medal of Freedom**, given to people who make the world better.

1952

After **graduating** from high school, Jane **takes a job** as a secretary.

1956

Jane's friend invites her to visit—and Jane saves money so she can **sail to Kenya, Africa**.

1957

Jane meets famous animal behavior scientist **Dr. Louis Leakey**, and becomes his assistant.

1960

Dr. Leakey chooses Jane to **study the chimpanzees**. Jane and her mother arrive in Gombe Stream Reserve in Tanzania.

1986

Jane becomes an **activist**, fighting for chimps' protection and survival.

1991

Jane founds **Roots & Shoots**, a global environmental action group for kids.

2018

On July 14, the first **World Chimpanzee Day** is celebrated to raise awareness of humans' closest relative.

2024

Jane enjoys her **90th birthday** with ninety dogs!